AF887996

Who Am I?

Coral Reef Animals

Written by Read With You
Center for Language Research and Development

The characters and events portrayed in this book are fictitious. Any similarity to real persons, living or dead, is coincidental and not intended by the author.

Published by Read With You Publishing

Designed by Read With You Center for Language Research and Development

ISBN-13: 978-1-944710-70-5
ISBN-10: 1-944710-70-1

Third Edition July 2019
Printed in the United States of America.

I have stripes on my body.

Who am I?

clown fish

I don't have eyes.

Who am I?

sea urchin

I have many pointy fins.

Who am I?

lionfish

I have eight arms.

Who am I?

octopus

I can regrow my arms.

Who am I?

starfish

I don't have a backbone.

Who am I?

shrimp

My dad gives birth to my brothers and sisters.

Who am I?

seahorse

My mouth is on the bottom side of my body.

Who am I?

stingray

Clown Fish

Sea Urchin

Octopus

Lionfish

I am a clown fish.

I am a sea urchin.

I am an octopus.

I am a lionfish.

We are coral reef animals.

Starfish

Shrimp

Seahorse

Stingray

I am a starfish.

I am a shrimp.

I am a seahorse.

I am a stingray.

The coral reefs are home to millions of animals like us.

A coral reef is a long line of coral in warm, shallow water.

Made in United States
Orlando, FL
07 November 2026

90337517R00017